Venus

The Veiled Twin

JD ARDEN

Preface: A Shrouded Mystery

Venus, the second planet from the Sun, is Earth's twin in many respects. Similar in size, mass, and composition, it has often been called our "sister planet." To early astronomers, Venus seemed like a mirror of Earth—a world that might harbor oceans, continents, and even life beneath its thick, reflective clouds. Its brilliant appearance, shining brightly as both the morning and evening star, inspired myths and stories that tied it to love, beauty, and perfection.

But as we have learned more about Venus, this romantic vision has faded. Beneath its shroud of clouds lies a world of extremes—a surface hot enough to melt lead, an atmosphere of crushing pressure and toxic gases, and a history of transformation that turned it from a potentially habitable world into one of the most hostile environments in the solar system.

Venus is Earth's twin, but it is also its opposite. Where Earth thrives with life, Venus seethes with acid rain and superheated winds. Where Earth's climate is stable and nurturing, Venus's is a cautionary tale of runaway processes that spiral out of control.

And yet, Venus remains a place of fascination. Its mysteries, from its volcanic surface to its lost oceans, challenge our understanding of planetary evolution. Its extremes push the boundaries of what we consider possible. And its lessons, particularly about climate and habitability, resonate far beyond its scorching atmosphere.

This book seeks to explore Venus not only as a physical world but as a symbol of transformation, resilience, and the unexpected paths planets can take. It is a journey into a veiled mystery, a search for understanding beneath the clouds, and a reflection on what Venus can teach us about Earth, life, and the universe.

Chapter 1: The Surface Beneath the Veil

For centuries, Venus guarded its surface in secrecy. Wrapped in an impenetrable layer of clouds, it defied even the most advanced telescopes, revealing only its dazzling brightness and steady motion across the sky. What lay beneath those clouds was a question that tantalized astronomers and inspired imaginations. Some envisioned lush jungles and vast oceans; others imagined barren deserts or volcanic plains.

The truth, as revealed by modern science, is stranger and more complex than anyone could have predicted. Beneath its thick atmosphere, Venus's surface is a world of volcanic expanses, fractured plains, and impact craters, shaped by processes that are both familiar and utterly alien. To uncover this hidden landscape, scientists turned to one of humanity's most transformative tools: radar mapping.

Radar, unlike visible light, can penetrate the dense clouds of Venus and bounce off its surface. The reflected signals, analyzed by instruments aboard spacecraft, have provided a detailed picture of Venus's terrain. The first comprehensive radar maps came from the Soviet Union's **Venera missions** in the 1970s and 1980s, which captured glimpses of the planet's surface during their brief but groundbreaking landings. These missions were followed by NASA's **Magellan spacecraft**, which orbited Venus in the early 1990s and created the most detailed maps of the planet to date.

What these missions revealed was a landscape dominated by volcanic activity. Venus is home to vast plains of solidified lava, broken by towering shield volcanoes and intricate networks of fissures and rifts. Some of these volcanoes, such as **Maat Mons**, rise over 8 kilometers above the surrounding plains, making them among the tallest mountains in the solar system. The prevalence of volcanic features suggests that Venus has been shaped by intense internal heat, driving eruptions that have resurfaced the planet repeatedly over millions of years.

One of the most intriguing aspects of Venus's surface is its relative youth. Radar data indicates that the planet's surface is only about 300 to 500 million years old, a mere fraction of the age of the solar system. This suggests that Venus underwent a global resurfacing event in its recent geological past, during which volcanic activity covered most of the planet in fresh lava. The cause of this event remains a mystery, but it hints at a world where geological processes operate on a massive and unpredictable scale.

Despite its volcanic nature, Venus lacks the plate tectonics that define Earth's geology. Instead of moving plates, Venus's crust appears to behave as a single, stagnant shell. Heat from the planet's interior builds up beneath this shell, eventually breaking through in cataclysmic eruptions. This stagnant lid tectonics contrasts sharply with Earth's dynamic plates, offering a glimpse into an alternative model of planetary evolution.

Another striking feature of Venus's surface is the **tesserae**, regions of highly deformed terrain characterized by intersecting ridges and valleys. These tesserae are some of the oldest parts of Venus's surface, predating the planet's volcanic plains. They are thought to hold clues about Venus's ancient history, possibly recording a time when the planet's conditions were more Earth-like.

Impact craters, though less common on Venus than on airless worlds like Mercury or the Moon, also provide valuable insights. The thick atmosphere of Venus protects its surface from smaller impacts, allowing only larger meteoroids to reach the ground. As a result, the planet's craters are relatively well-preserved and evenly distributed, offering a window into the processes that shape Venus's surface over time.

The exploration of Venus's surface has not been without challenges. The planet's extreme environment—temperatures of 475 degrees Celsius, pressures equivalent to those found 900 meters underwater on Earth, and corrosive sulfuric acid clouds—makes direct observation extraordinarily difficult. Landers that reach the surface survive only a few hours before succumbing to the harsh conditions. Yet, even these brief glimpses have provided invaluable data, from chemical analyses of the surface to high-resolution images of the rocky terrain.

The hidden surface of Venus is more than a geological curiosity. It is a record of a planet that has undergone profound changes, a testament to the forces that shape worlds and the extremes they can reach. By peeling back the veil of Venus's clouds, we uncover not just the details of its landscape but the story of a planet that might once have been Earth's twin and is now its alien opposite.

Venus's surface, with its volcanic plains and ancient tesserae, is a paradox—both a window into the planet's past and a reflection of its inhospitable present. It challenges our understanding of planetary geology and forces us to consider how planets evolve under extreme conditions. As we continue to map and study this veiled world, we are reminded that even the most familiar-looking planets can hold surprises, their true nature hidden just beneath the surface.

Chapter 2: The Runaway Greenhouse Effect

Venus is a world of superlatives, and none is more extreme than its atmosphere. With temperatures reaching 475 degrees Celsius and pressures over 90 times that of Earth, Venus's climate is the most hostile of any terrestrial planet in the solar system. At the heart of this inferno lies the **runaway greenhouse effect**, a process that transformed Venus from a potentially habitable planet into an unrelenting furnace. Understanding this phenomenon is not only key to unraveling Venus's history but also critical to comprehending the fragility of planetary climates, including our own.

The term "greenhouse effect" refers to the process by which a planet's atmosphere traps heat. On Earth, this effect is essential for life, maintaining temperatures that allow water to remain liquid. Gases like carbon dioxide (CO_2), water vapor, and methane absorb infrared radiation emitted by the planet's surface, preventing it from escaping into space. This trapped heat warms the surface, creating a stable climate conducive to life.

On Venus, however, this process spiraled out of control. The planet's thick atmosphere, composed of 96.5% carbon dioxide, acts as a thermal blanket, trapping nearly all the heat from the Sun. Unlike Earth, where water plays a moderating role, Venus's atmosphere has almost no water vapor. The small amounts that do exist are quickly broken apart by ultraviolet radiation, a process called photodissociation, leaving behind free hydrogen that escapes into space.

The absence of water exacerbates the greenhouse effect. Water vapor, when present in sufficient quantities, forms clouds that reflect sunlight, providing a cooling effect. On Venus, the lack of liquid water or a hydrological cycle removes this balancing mechanism, allowing the planet's atmosphere to absorb heat unchecked.

The surface temperature of Venus is hot enough to melt lead, and these temperatures are not localized but uniform across the planet, day and night, pole to equator. This is due to the incredible density of Venus's atmosphere, which efficiently redistributes heat through powerful convection currents. Even on the night side of the planet, the surface remains as scorching as on the sunlit side.

But Venus was not always this way. Evidence suggests that billions of years ago, Venus may have had oceans, a thinner atmosphere, and a climate more similar to Earth's. Early in the solar system's history, the Sun was cooler, emitting less energy. During this period, Venus might have harbored conditions suitable for liquid water and possibly life.

What changed? The runaway greenhouse effect likely began with a warming trend caused by an initial increase in solar radiation or volcanic outgassing of greenhouse gases like carbon dioxide. As temperatures rose, any surface water would have evaporated, increasing the concentration of water vapor in the atmosphere—a potent greenhouse gas. This feedback loop—more heat leading to more evaporation, leading to more heat—would have accelerated rapidly.

Once the oceans evaporated completely, Venus reached a tipping point. Without liquid water to absorb carbon dioxide from the atmosphere and form carbonate rocks, the planet lost one of its key regulatory mechanisms. On Earth, this process helps stabilize the climate over geological timescales, as carbon dioxide is sequestered into the crust. On Venus, the absence of this "carbon sink" allowed CO_2 to accumulate unchecked, amplifying the greenhouse effect.

The volcanic activity that dominates Venus's surface may have also played a role. Massive eruptions could have released vast quantities of carbon dioxide and sulfur dioxide into the atmosphere, further intensifying the greenhouse effect. Sulfur dioxide, a precursor to the planet's sulfuric acid clouds, contributes to the extreme conditions by creating a highly reflective upper atmosphere.

The result is an environment where temperatures and pressures are beyond anything humans could survive. The air at Venus's surface is so dense that it behaves more like a liquid than a gas, and the atmosphere's composition makes it corrosive, capable of eroding spacecraft in mere hours.

Venus

Understanding the runaway greenhouse effect on Venus is not just a study of planetary extremes; it is a cautionary tale. While Earth's climate is currently stable, it is not immune to feedback loops that could disrupt this balance. Human activities, particularly the release of greenhouse gases, have already begun to alter the planet's climate. Venus demonstrates how quickly a habitable world can transform when tipping points are crossed.

The lessons from Venus are stark but invaluable. They show us the importance of understanding and preserving the mechanisms that regulate climate, such as the carbon cycle and the role of water. They remind us that planets are dynamic systems, where small changes can lead to dramatic and irreversible shifts.

From a scientific perspective, Venus offers a unique laboratory for studying atmospheric physics, climate dynamics, and planetary evolution. Its thick atmosphere provides insights into how gases interact with heat, pressure, and solar radiation, while its history raises questions about the conditions that sustain or destroy habitability.

But Venus's runaway greenhouse effect is more than a scientific curiosity—it is a warning. It reminds us that planets, even those as seemingly stable as Earth, exist in a delicate balance. To study Venus is to glimpse a possible future, one shaped by the very forces that we are only beginning to understand and influence.

Venus, shrouded in clouds and wrapped in heat, is a testament to the power of planetary processes and the thin line between paradise and inferno. Its story is one of transformation—a world that began as a twin to Earth and became a cautionary tale, a veiled mirror of what might be.

Chapter 3: Acid Clouds and Winds

If the runaway greenhouse effect makes Venus the solar system's furnace, its atmosphere elevates it to a cauldron of toxic chaos. The planet's skies are thick with sulfuric acid clouds, whipped into violent storms by winds that rage at hundreds of kilometers per hour. Venus's atmosphere is a world unto itself—a dynamic, layered system of heat, chemicals, and movement, utterly alien and profoundly hostile.

The thick, reflective clouds of Venus are the planet's most distinctive feature, obscuring its surface and creating the illusion of a smooth, featureless world. These clouds, composed primarily of sulfuric acid droplets, form in the upper atmosphere, where sulfur dioxide and water vapor react under the influence of ultraviolet light. The process generates a toxic brew that rains acid onto the lower atmosphere, though the extreme heat near the surface ensures that these droplets evaporate long before reaching the ground.

The sulfuric acid clouds serve as both a shield and a trap. They reflect a significant portion of the Sun's light, giving Venus its brilliant, reflective appearance as seen from Earth. But beneath this reflective layer, the clouds trap immense amounts of heat, contributing to the planet's superheated surface temperatures.

These clouds are not static; they are part of a complex weather system driven by Venus's unique atmospheric dynamics. While the planet itself rotates extremely slowly—one Venusian day lasts 243 Earth days—its upper atmosphere moves at extraordinary speeds, circling the planet in just four Earth days. This phenomenon, known as **super-rotation**, is one of Venus's greatest mysteries.

Super-rotation creates winds that reach speeds of up to 360 kilometers per hour in the upper atmosphere, far faster than the rotation of the planet below. These winds drive massive storms, generating turbulence and waves that ripple through the clouds. Scientists are still unraveling the mechanisms behind this super-rotation, which may involve a combination of solar heating, thermal tides, and the planet's near lack of a magnetic field to slow atmospheric movement.

Venus

Beneath the raging winds of the upper atmosphere lies a series of increasingly dense and stable layers, each with its own unique properties. The upper layers, at altitudes of 50 to 70 kilometers, are relatively temperate, with temperatures ranging from 0 to 50 degrees Celsius. This region, sometimes called the **cloud deck**, is where sulfuric acid droplets form and where the super-rotation is most intense.

Descending through the atmosphere, the conditions become increasingly hostile. Temperatures and pressures rise sharply, creating a crushing environment where gases behave more like liquids. At the surface, the atmosphere is so dense that it exerts a pressure equivalent to being 900 meters underwater on Earth. In this infernal region, carbon dioxide dominates, with small amounts of nitrogen and trace gases completing the mix.

Among the most intriguing features of Venus's atmosphere are the mysterious **dark streaks** observed in the ultraviolet spectrum. These streaks, seen in the cloud tops, absorb UV light and create patterns that shift and evolve over time. The exact composition of these streaks remains unknown, though some scientists speculate they could be caused by sulfur compounds or other complex molecules. A more provocative hypothesis suggests that they could be signs of microbial life—extremophiles capable of surviving in the temperate cloud layers where conditions are relatively mild. While this idea remains speculative, it underscores the atmosphere's potential as a region of scientific discovery.

The violent weather on Venus also produces striking phenomena, including lightning and thunder. Observations from spacecraft like **Venera** and **Pioneer Venus** have detected electromagnetic discharges in the clouds, evidence of lightning storms within the sulfuric acid layer. These storms, while less frequent than those on Earth, contribute to the dynamic and chaotic nature of Venus's atmosphere.

Venus's winds and clouds also play a role in shaping its surface. Although the dense lower atmosphere limits direct erosion, the circulation of gases and particulates contributes to the redistribution of heat and the chemical weathering of surface materials. This interaction between the atmosphere and the surface creates a feedback loop that helps sustain the planet's extreme conditions.

Understanding Venus's atmospheric dynamics is more than an academic exercise; it has practical implications for planetary science and exploration. The super-rotation of Venus's atmosphere, for example, offers a natural laboratory for studying high-speed wind systems and their effects on planetary climates. Similarly, the chemistry of the sulfuric acid clouds provides insights into how volatile compounds behave under extreme conditions, with applications ranging from industrial processes to the search for life on other worlds.

From a philosophical perspective, Venus's atmosphere challenges our notions of stability and balance. Its clouds and winds operate on scales that dwarf anything on Earth, creating a system where chaos and order coexist. The sulfuric acid clouds, with their corrosive rain and volatile chemistry, remind us that beauty—such as Venus's bright appearance in the night sky—can conceal immense hostility.

Venus's atmosphere is a paradox, both reflective and destructive, both dynamic and unyielding. It is a system of extremes that defies simple explanations, a place where winds race faster than the planet itself and rains never reach the ground.

In the swirling acid clouds and relentless winds of Venus, we find a world that operates on its own terms, shaped by forces that challenge our understanding and push the boundaries of what we consider possible. Venus's atmosphere is not just a veil—it is a force of nature, a reminder of the universe's capacity for both creation and destruction, beauty and chaos.

Chapter 4: Venusian Evolution

Venus, Earth's so-called twin, is a planet defined by contrasts. While the two worlds share similar sizes, compositions, and origins, their evolutionary paths could not be more different. Today, Venus is a sweltering inferno, its surface temperatures high enough to melt lead, its atmosphere thick with carbon dioxide and corrosive clouds of sulfuric acid. Yet, Venus may not have always been this way. Billions of years ago, it could have been a temperate world, with oceans, a thinner atmosphere, and the potential to sustain life. Understanding how Venus changed from a planet with Earth-like possibilities to the most hostile environment in the solar system is a story of transformation and extremes.

In the early solar system, Venus likely began as a planet much like Earth. Positioned near the inner edge of the Sun's habitable zone, it may have had a thinner atmosphere and conditions that allowed liquid water to pool on its surface. Vast oceans might have covered Venus, and volcanic activity could have driven a dynamic climate. During this early period, the Sun was less luminous, and Venus's distance from its heat seemed compatible with habitability.

But proximity to the Sun comes with risks, and over time, Venus began to experience a gradual but profound shift. As the Sun aged, it brightened, emitting more energy and heating the planets in its orbit. For Venus, this increase in solar radiation initiated a chain reaction. As surface temperatures rose, water from its hypothetical oceans evaporated into the atmosphere. Water vapor, a powerful greenhouse gas, trapped more heat, leading to further evaporation. This positive feedback loop spiraled out of control, creating what is known as a runaway greenhouse effect.

As temperatures soared, Venus's oceans boiled away entirely. Without liquid water to absorb atmospheric carbon dioxide and lock it into rocks, volcanic emissions of CO_2 began to dominate. On Earth, this natural process is balanced by the carbon cycle, where oceans and biological activity help regulate atmospheric composition. On Venus, the absence of these mechanisms allowed CO_2 to accumulate unchecked, amplifying the greenhouse effect further.

At the same time, Venus's atmosphere became increasingly dense and hostile. As temperatures rose and water vapor filled the air, ultraviolet radiation from the Sun broke apart water molecules in a process called photodissociation. The lighter hydrogen atoms escaped into space, leaving behind oxygen that either bonded with surface materials or contributed to the atmosphere's growing instability. This loss of hydrogen ensured that water could never reform on Venus, locking the planet into a cycle of arid desolation.

Volcanism also played a pivotal role in Venus's transformation. The planet's surface is dominated by vast lava plains and towering shield volcanoes, evidence of intense and sustained geological activity. Volcanic eruptions released massive quantities of greenhouse gases, further thickening the atmosphere and feeding the runaway climate. At some point between 300 and 500 million years ago, Venus appears to have experienced a catastrophic global resurfacing event. During this time, volcanic activity likely buried much of the planet's older surface beneath fresh lava, erasing ancient craters and reshaping its landscape entirely.

Unlike Earth, Venus lacks the plate tectonics that regulate geological processes. Earth's moving tectonic plates allow heat to escape gradually from the planet's interior, maintaining a dynamic but balanced system. Venus, in contrast, has a single, stagnant crust. As heat builds up beneath this immobile surface, it is released sporadically in massive volcanic outbursts, leading to dramatic and unpredictable changes in the planet's surface and atmosphere.

The global resurfacing of Venus may have marked the final turning point in its evolutionary journey. The thickened atmosphere, laden with carbon dioxide and sulfur dioxide, created conditions that were unrelentingly hostile. The planet became a pressure cooker, its surface temperatures soaring to levels where metals like lead and zinc would melt.

Despite its current hostility, Venus's past as a potentially habitable world remains one of the most intriguing possibilities in planetary science. If Venus once had oceans, could it also have supported life? Did microbial organisms thrive in its ancient seas, only to perish as the planet heated beyond the limits of survival? These questions remain unanswered, but they fuel the ongoing exploration of Venus, as scientists search for evidence of its watery past in its surface features and atmospheric composition.

The story of Venus's evolution is not just about what was lost but about the delicate balance that makes planets habitable. Venus reveals how easily that balance can be disrupted. While its transformation occurred over billions of years, the processes that drove it—feedback loops, volcanic activity, and atmospheric instability—are not unique to Venus. Earth's climate, though currently stable, is also vulnerable to tipping points that could lead to dramatic shifts. Venus serves as both a warning and a lesson, showing how small changes in planetary conditions can lead to irreversible consequences.

Studying Venus's evolution also deepens our understanding of planetary diversity across the universe. As astronomers discover exoplanets orbiting distant stars, many of these worlds share characteristics with Venus, such as thick atmospheres or close proximity to their suns. Venus becomes a touchstone for understanding how planets in extreme environments behave, offering insights that extend far beyond our solar system.

Venus's journey from Earth's twin to an uninhabitable furnace is a story of contrasts, resilience, and transformation. It reminds us that planets are dynamic systems, shaped by their environments and internal processes in ways that can lead to vastly different outcomes. As we continue to study Venus, we are not just uncovering the history of a single planet but gaining a deeper appreciation for the forces that shape worlds—and for the fragile balance that makes Earth so uniquely suited to life.

Chapter 5: Venus in Myth and Art

Venus, the brightest celestial body in Earth's night sky after the Moon, has captivated humanity for millennia. Known as the "morning star" and the "evening star," its steady brilliance and predictable movements have made it a symbol of beauty, constancy, and mystery. Across cultures and centuries, Venus has inspired myths, stories, and works of art, reflecting humanity's fascination with the heavens and our enduring desire to find meaning in the cosmos.

In Roman mythology, Venus was the goddess of love, beauty, and fertility. She represented both physical allure and the generative forces of nature, a duality that mirrored the planet's luminous appearance and its hidden mysteries. The Romans named the planet after her, seeing in its radiance a reflection of her divine qualities. This association with Venus, the goddess, cemented the planet's role as a symbol of desire and idealized femininity in Western culture.

The Greeks, who called the planet **Phosphoros** (the bringer of light) when it appeared in the morning and **Hesperos** (the evening star) when it shone at dusk, similarly tied it to their pantheon. They later merged these identities under the goddess **Aphrodite**, Venus's Greek counterpart. Aphrodite, like Venus, embodied love and beauty, but she also held associations with chaos and transformation, themes that resonate with the planet's extremes and its hidden, fiery surface.

Beyond the Mediterranean, Venus occupied a prominent place in other cultures as well. The Babylonians called the planet **Ishtar**, naming it after their goddess of love, war, and fertility. Ishtar's duality—as both a nurturing force and a warrior goddess—reflects the paradoxical nature of Venus: a brilliant, beautiful object that conceals a harsh and unyielding reality. Similarly, the ancient Maya associated Venus with warfare and used its cycles to time military campaigns, seeing its appearance in the sky as an omen of conflict.

In Hindu mythology, Venus is known as **Shukra**, a name derived from the teacher of the demons in Vedic tradition. Shukra is associated with wisdom, wealth, and the pursuit of worldly pleasures, embodying the planet's dual nature as both a beacon of enlightenment and a symbol of material desire.

In these myths and traditions, Venus is often tied to cycles, transitions, and duality. Its dual role as the morning and evening star made it a symbol of beginnings and endings, of continuity and change. These themes are reflected in art, literature, and symbolism throughout human history.

The Renaissance brought a resurgence of interest in Venus, both as a goddess and a planet. Artists like Sandro Botticelli immortalized her in works like **The Birth of Venus**, where she emerges from the sea as an idealized figure of beauty and creation. This image, though rooted in mythology, also reflects the Renaissance fascination with celestial harmony and the human connection to the cosmos. Venus became a muse not just for painters but for poets, musicians, and astronomers, each interpreting its significance through their unique lens.

In literature, Venus often serves as a symbol of unattainable beauty or divine inspiration. Dante referenced the planet in his **Divine Comedy**, placing it in the sphere of love within his celestial hierarchy. For writers like Shakespeare, who mentioned Venus in works like **Venus and Adonis**, the planet embodied themes of passion and longing, tying its celestial brightness to human emotion.

Science, too, found inspiration in Venus. Early astronomers like Galileo Galilei marveled at its phases, visible through a telescope, which provided crucial evidence for the heliocentric model of the solar system. Venus's predictable cycles and its role as a timekeeper in ancient calendars linked it to humanity's efforts to understand and measure the cosmos, blending myth and observation.

But Venus's symbolic role is not limited to the past. In modern times, its mystique endures, even as science reveals the planet's inhospitable nature. Its paradoxical identity—a brilliant light in the sky hiding a searing, uninhabitable surface—makes it a powerful metaphor for the dualities of existence. It is both beautiful and destructive, inviting and forbidding, a reminder that appearances can conceal profound complexity.

Venus

Artists and writers today continue to draw from Venus's rich legacy. Science fiction authors have imagined Venus as both a utopian paradise and a cautionary tale of planetary climate catastrophe. Filmmakers, painters, and poets have used its name and image to explore themes of love, power, and transformation, ensuring that the planet's cultural significance evolves alongside our understanding of it.

Venus, as a symbol, transcends time and culture. Its light has guided ancient warriors and inspired Renaissance artists, its myths have shaped divine archetypes, and its enigmatic nature continues to challenge our understanding of the universe. It reminds us that even in the vastness of space, the cosmos reflects our own struggles, dreams, and contradictions.

The planet Venus is more than a celestial body; it is a mirror for humanity's enduring fascination with beauty, love, and mystery. Its myths and art tell a story not only of the heavens but of ourselves, revealing how deeply we are connected to the worlds that shine above us.

Chapter 6: Explorations of Venus

Venus, shrouded in clouds and blazing with heat, has long been one of the most enigmatic worlds in the solar system. Its brightness and proximity made it an object of fascination for early astronomers, yet its hostile environment proved a daunting challenge for direct exploration. Over the decades, humanity's attempts to study Venus have ranged from flybys to ambitious landings, each mission uncovering new insights while raising more questions. The exploration of Venus is a story of determination, ingenuity, and the relentless pursuit of knowledge.

The first successful mission to Venus came in 1962, when NASA's **Mariner 2** became the first spacecraft to perform a close flyby of another planet. Mariner 2 revealed Venus's searing surface temperatures and lack of a significant magnetic field, confirming that its atmosphere was vastly different from Earth's. These findings were a sobering reality check for those who had once imagined Venus as a lush, Earth-like world.

The Soviet Union, however, was the first to make direct contact with Venus's surface. The **Venera program**, spanning from the 1960s to the 1980s, marked a series of pioneering efforts to explore the planet. The early Venera missions encountered numerous challenges, with spacecraft succumbing to Venus's crushing atmospheric pressure, extreme heat, and corrosive clouds. Yet, with each failure, engineers refined their designs, and the Venera program achieved several historic milestones.

Venera 7, launched in 1970, became the first spacecraft to transmit data from the surface of another planet. Though it survived for only 23 minutes, it provided critical information about Venus's pressure, temperature, and atmospheric composition. Subsequent missions, such as **Venera 9** and **Venera 10**, captured the first images of Venus's rocky terrain, revealing a landscape of fractured plains and scattered boulders.

One of the most significant achievements of the Venera program came with **Venera 13** and **Venera 14** in 1982. These landers not only transmitted images of the surface but also conducted soil analyses, finding a composition similar to basaltic rocks on Earth. They confirmed that Venus's surface was shaped by volcanic activity, laying the groundwork for our current understanding of the planet's geology.

Venus

The United States returned to Venus with NASA's **Pioneer Venus** program in the late 1970s. The **Pioneer Venus Orbiter** and a suite of atmospheric probes provided detailed data about the planet's upper atmosphere, cloud structure, and circulation patterns. These missions revealed the existence of the mysterious super-rotation of Venus's upper atmosphere, where winds travel far faster than the planet itself rotates.

In 1990, NASA's **Magellan** spacecraft revolutionized our understanding of Venus's surface. Using radar mapping, Magellan penetrated the planet's dense clouds to create detailed maps of its terrain. The spacecraft revealed vast volcanic plains, towering shield volcanoes, and the enigmatic tesserae—regions of highly deformed terrain that hint at the planet's ancient past. Magellan's data confirmed that Venus's surface was geologically young, resurfaced by volcanic activity within the last 300–500 million years.

Despite these successes, Venus exploration has remained sporadic. While missions like the European Space Agency's **Venus Express** (2005–2014) and Japan's **Akatsuki** orbiter (launched in 2010) have provided valuable insights into Venus's atmosphere and climate, no spacecraft has landed on Venus since the Venera program. The challenges of exploring Venus's surface—its intense heat, pressure, and corrosive atmosphere—continue to limit the scope of missions.

Yet, interest in Venus is resurging. NASA's planned **VERITAS** mission, set to launch in the 2030s, aims to use advanced radar to map Venus's surface in unprecedented detail, investigating its geological history and active processes. Similarly, the European Space Agency's **EnVision** mission will study the planet's atmosphere, surface, and interior, searching for clues about its evolution and the potential for past habitability. These missions represent a new chapter in Venus exploration, driven by the recognition of its importance in understanding planetary climates and evolution.

The lessons learned from Venus are not confined to its own mysteries. Studying Venus helps scientists refine their models of planetary atmospheres, geological processes, and climate dynamics. It provides a basis for understanding exoplanets, particularly the many "hot Earths" discovered orbiting close to their stars. Venus, with its extremes, becomes a case study in how planets evolve under intense solar radiation and atmospheric pressures.

The exploration of Venus also holds philosophical significance. It challenges humanity to confront the limits of what we can endure and achieve. Venus, with its dense clouds and unforgiving surface, demands ingenuity and resilience from those who seek to study it. Each mission to Venus is a testament to human curiosity, a drive to understand not just what lies within our reach but what exists at the edge of possibility.

While Venus is unlikely to host life as we know it, the search for organic compounds and signs of past habitability continues to drive exploration. Recent detections of **phosphine**, a gas that could indicate microbial life in the planet's upper atmosphere, have reignited debates about Venus's potential for harboring life in its temperate cloud layers. Although the findings remain contentious, they underscore the fact that Venus still holds secrets waiting to be uncovered.

In many ways, Venus is a reminder of the unpredictability of discovery. From the early flybys to the modern radar maps, every mission to Venus has revealed something unexpected, challenging our preconceptions and deepening our understanding of the solar system. The exploration of Venus is far from over, and as technology advances, so too will our ability to peel back the layers of its veiled world.

Venus's story is one of endurance, transformation, and revelation. It is a planet that defies easy answers, a world that both mirrors and contrasts with Earth. As we continue to explore Venus, we are not only uncovering its mysteries but also expanding the boundaries of human knowledge, pushing ever closer to understanding the forces that shape planets and the possibilities that lie beyond.

Chapter 7: The Dangers of Idealization

Venus has long been a symbol of beauty, love, and perfection. Its brilliant light, steady movements, and prominent place in the sky inspired cultures across the world to associate it with the divine, the ideal, and the desirable. From Venus the goddess of Roman mythology to the shining "morning star" of Biblical texts, the planet has been celebrated as an emblem of harmony and allure. Yet, beneath this idealized image lies a world of chaos, extremes, and destruction—a fiery inferno concealed by an outward veil of radiant clouds.

This contrast between perception and reality is not just a curiosity of planetary science; it is a profound metaphor for human tendencies to idealize what appears beautiful while ignoring underlying complexities. Venus challenges our assumptions about equating outward beauty with goodness, harmony, or benevolence. Its nature reminds us that the allure of perfection often conceals contradictions, and that our ideals, when unexamined, can lead to misunderstanding or even peril.

For much of human history, Venus's brightness and apparent constancy reinforced its reputation as a celestial symbol of balance. Early astronomers noted its predictable cycles, its presence as both the morning and evening star, and its role in marking agricultural and ceremonial calendars. Its association with deities of love and fertility further rooted Venus in the human psyche as a harbinger of life and vitality.

Yet this idealized view began to unravel with the advent of modern science. The first telescopic observations revealed Venus's phases, much like those of the Moon, proving that it orbited the Sun rather than Earth—a revelation that challenged centuries of geocentric thought. As astronomers continued to study the planet, they discovered that Venus was not the paradise many had imagined. Instead of oceans and lush landscapes, Venus's surface was revealed to be blisteringly hot, its atmosphere suffocatingly dense and toxic.

Venus

The transformation of Venus from an idyllic twin of Earth to a cautionary tale of extremes underscores the risks of romanticizing the unknown. Early astronomers, guided more by imagination than evidence, projected their hopes and ideals onto Venus. These projections, though inspiring, were ultimately misleading. They remind us that idealization, while comforting, can blind us to reality.

This tendency to idealize is not confined to Venus or even to astronomy. It is a human trait that manifests in how we view beauty, relationships, power, and progress. We often assume that what is outwardly appealing must also be inherently good, that what seems harmonious must be free of conflict. Venus's story exposes the fallacy of this assumption. Its clouds, so brilliant and reflective, are composed of sulfuric acid, hiding an uninhabitable landscape of volcanic plains and crushing pressure. Beneath the appearance of perfection lies a reality that defies expectation.

In the context of planetary science, this realization has profound implications. The discovery of Venus's hostile environment challenged early models of habitability and forced scientists to reconsider the criteria for what makes a planet "Earth-like." It revealed the importance of looking beyond surface similarities to understand the deeper dynamics that shape a world.

The same lesson applies to how we approach ideals in human life. The pursuit of beauty, love, or perfection can inspire greatness, but it can also lead to disillusionment if we fail to recognize the complexities and imperfections that lie beneath. Venus, both as a planet and as a symbol, serves as a reminder that the most beautiful things are often the most complex, and that understanding requires looking beyond appearances.

This is not to say that idealization is without value. The myths and stories surrounding Venus have inspired art, poetry, and exploration for centuries. They have connected humanity to the cosmos and fostered a sense of wonder that drives discovery. Even today, the image of Venus as the morning star—shining bright in the dawn sky—carries a sense of hope and possibility.

But to truly appreciate Venus, we must embrace its contradictions. Its beauty and hostility, its brightness and opacity, its role as both a symbol

Venus

of love and a cautionary tale of extremes—all these facets are part of its story. By accepting the complexities of Venus, we move beyond mere idealization to a deeper understanding, one that honors the planet for what it is rather than what we imagine it to be.

Venus teaches us that perfection is an illusion, and that beauty often comes with layers of complexity and contradiction. It challenges us to approach the world—not just planets, but people, ideas, and systems—with curiosity and humility. To idealize is human, but to understand is greater.

As we continue to explore Venus, both scientifically and symbolically, we are reminded that the universe is not a reflection of our ideals but a reality that exists in its own right. Venus, in its dual nature, offers a lesson in seeing the world as it is, and in finding wonder not just in perfection but in the truths that lie beneath.

Chapter 8: Venus's Lessons for Earth

Venus, Earth's closest planetary neighbor and its so-called twin, stands as a stark warning. Despite its similar size, mass, and composition, Venus has become an uninhabitable inferno, with surface temperatures hot enough to melt lead, an atmosphere saturated with carbon dioxide, and a crushing pressure equivalent to being nearly a kilometer underwater. Yet billions of years ago, Venus may have been more like Earth, possibly with oceans, a temperate climate, and conditions suitable for life.

What changed? And what does Venus's history teach us about the fragility of habitability? In studying Venus, we uncover lessons not just about its evolution but about the precarious balance that sustains life on Earth. These lessons are as urgent as they are profound, offering insights into the dynamics of planetary climates and the potential consequences of tipping points that lead to irreversible change.

At the heart of Venus's transformation lies the **runaway greenhouse effect**, a process where rising temperatures lead to feedback loops that accelerate warming. On Venus, this began with its slightly closer proximity to the Sun. Over time, increased solar radiation caused its surface temperatures to rise, evaporating any liquid water that may have existed. Water vapor, a potent greenhouse gas, further trapped heat in the atmosphere, leading to more evaporation and even higher temperatures.

This feedback loop spiraled out of control. As Venus's oceans evaporated entirely, the planet lost one of its primary mechanisms for regulating carbon dioxide. On Earth, water absorbs CO_2 from the atmosphere, allowing it to be stored in the oceans or locked away in carbonate rocks. Without this process, Venus's volcanic emissions of CO_2 accumulated unchecked, thickening the atmosphere and amplifying the greenhouse effect.

The parallels between Venus's runaway greenhouse effect and Earth's current climate crisis are impossible to ignore. Human activities—particularly the burning of fossil fuels—are releasing vast quantities of greenhouse gases into Earth's atmosphere, raising global temperatures and disrupting climatic stability. While Earth is far from the extremes of Venus, the processes driving its climate change are eerily similar, albeit on a smaller scale and much shorter timespan.

Venus serves as a cautionary tale, showing how quickly a planet's climate can tip into a state of no return. On Earth, scientists warn of similar tipping points: the melting of polar ice caps, the release of methane from thawing permafrost, the disruption of ocean currents, and the deforestation of rainforests, all of which could accelerate global warming beyond human control. Venus demonstrates the catastrophic consequences of such runaway processes.

But Venus's lessons are not solely about warnings. They also offer insights into the mechanisms that sustain and disrupt planetary climates. Understanding the dynamics of Venus's atmosphere, for example, helps scientists refine models of Earth's climate, improving predictions of how it will respond to human influence. The study of Venus also provides clues about how greenhouse gases interact with heat, how clouds affect planetary albedo, and how volcanic activity influences atmospheric composition—all factors that play critical roles in Earth's climate system.

Venus also highlights the importance of **feedback loops** in planetary evolution. On Earth, feedback loops can either stabilize or destabilize the climate. For example, ice and snow reflect sunlight, helping to cool the planet. As ice melts due to rising temperatures, less sunlight is reflected, and more is absorbed, leading to further warming. Similarly, Venus's loss of water vapor allowed more CO_2 to dominate its atmosphere, locking the planet into its current state. Recognizing these loops and their potential to cascade is essential for understanding and mitigating climate change on Earth.

The study of Venus has also deepened our understanding of the long-term evolution of planetary climates. Venus's current state is not just a result of its proximity to the Sun but a combination of factors, including its lack of a magnetic field, its geological history, and its loss of water. Each of these elements offers insights into how planets transition from one state to another and how fragile the conditions for habitability can be.

Venus

For Earth, these lessons underscore the importance of preserving the delicate balance that allows life to thrive. Our planet's climate is regulated by an intricate interplay of atmospheric, geological, and biological processes, from the carbon cycle to ocean circulation to the reflective power of ice caps. Disrupting any one of these systems risks setting off a chain reaction that could fundamentally alter the planet's environment.

Venus also serves as a reminder of humanity's role as stewards of Earth. Unlike Venus, which succumbed to natural processes, Earth's current trajectory is being shaped by human activity. The choices we make now—how we produce energy, manage resources, and address carbon emissions—will determine whether Earth remains a haven for life or edges closer to the extremes of its planetary twin.

From a philosophical perspective, Venus challenges us to confront the fragility of habitability and the interconnectedness of planetary systems. It forces us to consider how small changes—whether in planetary positioning or human activity—can lead to profound consequences over time. Venus is both a mirror and a warning, reflecting the potential fate of a world pushed beyond its limits.

Venus's lessons extend beyond Earth. As scientists search for habitable planets around distant stars, the study of Venus offers a template for understanding the conditions that make a world livable—or unlivable. Many exoplanets discovered so far are "hot Earths," rocky planets orbiting close to their stars. Venus provides a window into the processes that shape such worlds, offering clues about their climates, histories, and potential for life.

Ultimately, Venus teaches us about the boundaries of possibility. It reminds us that the conditions for life are not guaranteed, that planets are dynamic systems shaped by forces both internal and external. Venus, in its extremes, underscores the urgency of understanding our own planet and the fragility of the balance that sustains it.

The story of Venus is not just a cautionary tale but a call to action—a reminder that the choices we make today will shape the future of Earth and the generations that inherit it. It is a challenge to learn from the

Venus

cosmos and to use that knowledge to preserve the only habitable world we know.

Conclusion: The Twin That Wasn't

Venus and Earth began as twins, forged from the same cosmic dust, positioned close to the Sun in its habitable zone, and endowed with similar sizes, compositions, and potential. Yet, their destinies diverged in profound and dramatic ways. Earth flourished into a cradle for life, its climate stabilized by oceans and a dynamic atmosphere. Venus, by contrast, became a hostile inferno, its once-potentially temperate climate consumed by a runaway greenhouse effect.

The comparison between Venus and Earth is a tale of what could have been—and a stark reminder of how thin the line between habitability and hostility can be. Venus, once imagined as a lush paradise hidden beneath clouds, is now understood as a cautionary tale of extremes, a twin in size but an alien in character. Its evolution from a world of potential to one of unyielding heat and pressure provides profound lessons for understanding planetary systems, climate dynamics, and the fragility of life's conditions.

Venus's fate was shaped by a combination of factors, beginning with its proximity to the Sun. Slightly closer to the Sun than Earth, Venus received more solar energy, which set the stage for the runaway feedback loops that drove its transformation. As temperatures rose, Venus's oceans—if they existed—evaporated, flooding the atmosphere with water vapor. This potent greenhouse gas trapped more heat, accelerating the warming process and boiling away the last vestiges of liquid water.

The absence of water sealed Venus's fate. Without oceans to absorb carbon dioxide and lock it into the planet's crust, CO_2 accumulated in the atmosphere unchecked. Volcanic activity added even more greenhouse gases, creating a thick, oppressive atmosphere that now exerts 90 times the pressure of Earth's. As heat built up and water vapor was broken apart by ultraviolet radiation, Venus lost its hydrogen to space, ensuring that water could never reform.

The result is a planet dominated by extremes. Its surface temperatures exceed 475 degrees Celsius, its winds race at hundreds of kilometers per hour, and its clouds rain sulfuric acid. These conditions, unrelenting and alien, render Venus one of the least hospitable places in the solar system, a far cry from the Eden once imagined by early astronomers and poets.

Despite its extremes, Venus remains a crucial object of study, offering invaluable insights into the processes that shape planets. It challenges us to understand the factors that lead to climate stability or collapse, and it forces us to grapple with the fragility of habitability. Venus shows us that the conditions for life are not inevitable but precarious, reliant on a delicate balance of atmospheric, geological, and solar dynamics.

For Earth, Venus is both a mirror and a warning. The parallels between the two planets highlight the shared forces that govern their climates: the interplay of greenhouse gases, solar energy, and atmospheric feedback loops. While Earth's climate is currently stable, it is not immune to tipping points. The runaway processes that boiled Venus's oceans and thickened its atmosphere are reminders of what could happen if Earth's climate systems are pushed beyond their limits.

Venus also challenges our assumptions about planetary evolution and habitability. For decades, it was seen as Earth's twin, a potential second home in the cosmos. Its eventual reality—an uninhabitable furnace—forces us to reconsider how we evaluate other planets, both within our solar system and beyond. The lessons of Venus resonate in the search for exoplanets, helping scientists refine the criteria for what makes a world truly Earth-like.

Yet, Venus's story is not just one of loss. It is also a testament to the resilience of planetary systems and the creativity of nature. Venus has withstood billions of years of solar radiation, volcanic upheaval, and atmospheric transformation. Its surface, though hostile, is a record of geological processes that continue to shape its landscape. Its atmosphere, though toxic, is a dynamic system that challenges our understanding of chemistry and physics.

Philosophically, Venus invites reflection on the nature of beauty, balance, and potential. Its radiant clouds, visible as the brightest object in the night sky after the Moon, conceal an unrelenting inferno. This duality—between appearance and reality, between promise and peril—mirrors the complexity of the universe itself. Venus teaches us that beauty does not always equate to harmony and that potential, when unfulfilled, can lead to extremes.

As humanity continues to explore Venus, its lessons grow ever more urgent. The planet's history is a stark reminder of the consequences of climate change and the importance of preserving Earth's fragile balance. Venus is not just a distant world but a lens through which we can better understand our own. It challenges us to act with foresight, to recognize the interconnectedness of planetary systems, and to appreciate the delicate conditions that sustain life.

Venus is the twin that wasn't—a world that began with the same promise as Earth but ended in chaos. Yet, in its extremes, it offers insights, warnings, and inspiration. Its clouds and winds, its volcanic plains and tesserae, its paradoxical beauty and unyielding hostility—all these elements are part of a story that transcends its surface.

The tale of Venus is not just the story of a planet; it is a reflection of the forces that shape worlds, the fragility of habitability, and the enduring quest to understand our place in the cosmos. As we look to Venus, we see both a warning of what can go wrong and a wonder of what exists beyond our understanding.

End Note: Venus and Beyond

Venus is a world that defies expectations. Once imagined as a lush twin to Earth, it revealed itself to be an unrelenting furnace, its surface hidden beneath clouds of toxic sulfuric acid. Yet, Venus's story is not just about what was lost—it is about what we have gained through its study. The exploration of Venus has deepened our understanding of planetary evolution, climate dynamics, and the forces that govern worlds both familiar and alien.

For scientists, Venus is a case study in extremes. Its atmosphere, surface, and geological history offer a wealth of data about how planets change over time. From the runaway greenhouse effect that transformed its climate to the volcanic activity that shaped its surface, Venus demonstrates the delicate interplay between internal and external forces that determine a planet's fate. Its lessons resonate far beyond its blistering skies, offering insights into Earth's own climate and the conditions that might shape distant exoplanets.

As humanity seeks to understand the cosmos, Venus serves as both a challenge and an inspiration. Its extremes test the limits of our technology, pushing spacecraft to survive crushing pressures and searing temperatures. Each mission to Venus is an exercise in resilience, a reminder of the ingenuity required to explore worlds beyond our own.

The resurgence of interest in Venus marks a new chapter in planetary exploration. Missions like NASA's **VERITAS** and ESA's **EnVision** promise to uncover new details about Venus's surface, atmosphere, and interior, shedding light on its geological history and the processes that continue to shape it. These missions, equipped with advanced instruments, will help answer lingering questions about Venus's past and its potential for habitability.

Yet Venus is not just a scientific puzzle; it is a philosophical mirror. It challenges us to confront the fragility of habitability and the consequences of environmental imbalance. Its extremes force us to consider the boundaries of possibility and the resilience of worlds shaped by forces beyond human comprehension.

Venus

Venus also reminds us of the interconnectedness of planetary systems. Its transformation from a potentially habitable world to an unyielding furnace underscores the importance of understanding climate dynamics, both on Earth and elsewhere. Venus is a cautionary tale, but it is also a guide, offering insights that can help us protect our own planet and make informed decisions about its future.

As we look to the stars, Venus serves as a touchstone for understanding exoplanets and their potential for life. Many of the rocky worlds discovered orbiting distant stars share characteristics with Venus, including thick atmospheres and proximity to their suns. Studying Venus allows us to refine our models of these distant worlds, placing our solar system in a broader cosmic context.

But Venus's significance extends beyond science. It is a symbol of both beauty and complexity, a world that has inspired myths, art, and exploration for centuries. Its radiant appearance in the night sky continues to captivate, even as we uncover the harsh realities beneath its clouds. Venus teaches us to embrace dualities—to see both the allure of its brilliance and the stark truths it hides.

Venus is more than Earth's twin; it is a reminder of the universe's capacity for both creation and destruction, balance and chaos. It is a world of lessons—scientific, philosophical, and existential—that challenge us to look beyond appearances and seek deeper understanding.

As we continue to explore Venus, we carry its lessons forward, not just to other planets but to our own. In its veiled surface and fiery heart, Venus holds the keys to questions about life, habitability, and the resilience of worlds. Its story is one of transformation, endurance, and the relentless pursuit of knowledge—a story that mirrors our own journey as explorers of the cosmos.